Heatley Noble

A List of European Birds

Heatley Noble

A List of European Birds

ISBN/EAN: 9783743313507

Manufactured in Europe, USA, Canada, Australia, Japa

Cover: Foto ©berggeist007 / pixelio.de

Manufactured and distributed by brebook publishing software (www.brebook.com)

Heatley Noble

A List of European Birds

A LIST

OF

EUROPEAN BIRDS,

INCLUDING ALL THOSE FOUND IN THE

WESTERN PALÆARCTIC AREA,

WITH A

SUPPLEMENT

Containing species said to have occurred, but which, for various reasons, are inadmissible.

BY

HEATLEY NOBLE, F.Z.S.

LONDON :

R. H. PORTER,

7, Princes Street, Cavendish Square, W.

1898.

INTRODUCTION.

As there is no list of European Birds up to date, it is hoped that the following may be of use to collectors, for reference and exchange of skins and eggs, also for labelling specimens, especially as the Supplement contains a list of those birds of disputed occurrence, printed in the *same type*, a feature not found in any other work of the kind.

My thanks are due to many friends for their kind assistance, and especially to Mr. Dresser, for permitting me to use his "List of European Birds."

HEATLEY NOBLE.

Temple Combe, Henley-on-Thames,
July, 1898.

A LIST OF EUROPEAN BIRDS.

AVES CARINATÆ.
ÆGITHOGNATHÆ.
Order I. PASSERES.
Suborder *OSCINES*.

Section 1. **OSCINES DENTIROSTRES.**

Family TURDIDÆ.

Subfamily TURDINÆ.

1. *Turdus viscivorus*, Linn.
   Missel-Thrush.

2. *Turdus musicus*, Linn.
   Song-Thrush.

3. *Turdus iliacus*, Linn.
   Redwing.

4. *Turdus pilaris*, Linn.
   Fieldfare.

5. *Turdus naumanni*, Temm.
   Red-tailed Fieldfare.

6. *Turdus dubius*, Bechst.
   Dusky Thrush.

7. *Turdus ruficollis*, Pall.
    Red-throated Thrush.

8. *Turdus obscurus*, Gmel.
    Pale Thrush.

9. *Turdus varius*, Pall.
    White's Thrush.

10. *Turdus atrigularis*, Temm.
    Black-throated Thrush.

11. *Turdus sibiricus*, Pall.
    Siberian Thrush.

12. *Turdus merula*, Linn.
    Blackbird.

13. *Turdus torquatus*, Linn.
    Ring-Ouzel.

14. *Turdus alpestris*, (C. L. Brehm).
    Alpine Ring-Ouzel.

15. *Turdus pallasi*, Cab.
    Hermit Thrush.

16. *Turdus swainsoni*, Cab.
    Olive-backed Thrush.

17. *Monticola saxatilis* (Linn.).
    Rock-Thrush.

18. *Monticola cyanus* (Linn.).
    Blue Rock-Thrush.

Subfamily CINCLINÆ.

19. *Cinclus aquaticus*, Bechst.
    Common Dipper.

20. *Cinclus melanogaster*, C. L. Brehm.
    Black-bellied Dipper.

21. *Cinclus albicollis* (Vieill.).
Pale-backed Dipper.

22. *Cinclus cashmiriensis*, Gould.
White-breasted Asiatic Dipper.

Subfamily SAXICOLINÆ.

23. *Saxicola œnanthe* (Linn.).
Common Wheatear.

24. *Saxicola seebohmi*, C. Dixon.
Seebohm's Wheatear.

25. *Saxicola isabellina*, Rüpp.
Isabelline Wheatear.

26. *Saxicola albicollis*, Vieill.
Black-eared Wheatear.

27. *Saxicola vittata*, Hemp & Ehr.
Ehrenberg's Wheatear.

28. *Saxicola stapazina*, Vieill.
Russet Wheatear.

29. *Saxicola melanoleuca* (Güld.).
Black-throated Wheatear.

30. *Saxicola deserti*, Rüpp.
Desert-Wheatear.

31. *Saxicola finschi*, Heugl.
Arabian Wheatear.

32. *Saxicola mœsta*, Licht.
Tristram's Wheatear.

33. *Saxicola xanthoprymna*, Ehr.
Red-rumped Wheatear.

34. *Saxicola chrysopygia*, De Filippi.
Red-tailed Wheatear.

35. *Saxicola lugens*, Licht.
 Pied Wheatear.

36. *Saxicola morio*, Ehr.
 Eastern Pied Wheatear.

37. *Saxicola monacha*, Temm.
 Hooded Wheatear.

38. *Saxicola leucopyga*, C. L. Brehm,
 White-rumped Wheatear.

39. *Saxicola albinigra*, Hume.
 Hume's Wheatear.

40. *Saxicola leucura* (Gm.).
 Black Wheatear.

41. *Saxicola picata*, Blyth.
 Black-backed Wheatear.

42. *Pratincola rubetra* (Linn.).
 Whinchat.

43. *Pratincola rubicola* (Linn.).
 Stonechat.

44. *Pratincola dacotæ*, Meade-Waldo.
 Canarian Stonechat.

45. *Pratincola maura* (Pall.).
 Eastern Stonechat.

46. *Pratincola hemprichi* (Ehr.).
 White-tailed Stonechat.

47. *Pratincola caprata* (Linn.).
 Pied Stonechat.

48. *Ruticilla phœnicurus* (Linn.).
 Redstart.

49. *Ruticilla mesoleuca* (Ehr.).
 Ehrenberg's Redstart.

50. *Ruticilla rufiventris* (Vieill.).
    Indian Redstart.

51. *Ruticilla titys* (Scop.).
    Black Redstart.

52. *Ruticilla ochrura* (Gmel.).
    Gould's Redstart.

53. *Ruticilla moussieri* (Olph-Gall.).
    Moussier's Redstart.

54. *Ruticilla erythrogastra* (Güld.).
    Güldenstädt's Redstart.

55. *Ruticilla erythronota* (Eversm.).
    Eversmann's Redstart.

Subfamily SYLVIINÆ.

56. *Cyanecula wolfi*, C. L. Brehm.
    White-spotted Bluethroat.

57. *Cyanecula suecica* (Linn.).
    Red-spotted Bluethroat.

58. *Erithacus rubecula* (Linn.).
    Redbreast.

59. *Erithacus hyrcanus*, Blanf.
    Persian Redbreast.

60. *Calliope camtschatkensis* (Gm.).
    Ruby-throated Warbler.

61. *Cossypha gutturalis*, Guérin.
    White-throated Robin-Chat.

62. *Nemura cyanura* (Pall.).
    Red-flanked Bluetail.

63. *Daulias luscinia* (Linn.).
    Nightingale.

64. *Daulias philomela* (Bechst.).
 Northern Nightingale.

65. *Daulias hafizi* (Severtz.).
 Persian Nightingale.

66. *Sylvia cinerea*, Bechst.
 Whitethroat.

67. *Sylvia curruca* (Linn.).
 Lesser Whitethroat.

68. *Sylvia minuscula*, Hume.
 Least Whitethroat.

69. *Sylvia althæa*, Hume.
 Himalayan Whitethroat.

70. *Sylvia subalpina*, Bonelli.
 Subalpine Warbler.

71. *Sylvia mystacea* (Ménétr.).
 Ménétries' Warbler.

72. *Sylvia conspicillata*, Marm.
 Spectacled Warbler.

73. *Sylvia nana* (Hemp. & Ehr.).
 Desert-Warbler.

74. *Sylvia melanothorax*, Tristram.
 Palestine Warbler.

75. *Sylvia melanocephala* (Gm.).
 Sardinian Warbler.

76. *Sylvia momus* (Ehr.).
 Bowman's Warbler.

77. *Sylvia orphea*, Temm.
 Orphean Warbler.

78. *Sylvia rueppelli*, Temm.
 Rüppell's Warbler.

79. *Sylvia atricapilla* (Linn.).
Blackcap.

80. *Sylvia hortensis*, Bechst.
Garden-Warbler.

81. *Sylvia nisoria*, Bechst.
Barred Warbler.

82. *Melizophilus undatus* (Bodd.).
Dartford Warbler.

83. *Melizophilus sardus* (Marm.).
Marmora's Warbler.

⤴ 84. *Melizophilus deserticola* (Trist.).
Tristram's Warbler.

Subfamily PHYLLOSCOPINÆ.

85. *Regulus cristatus*, Koch.
Golden-crested Wren.

86. *Regulus teneriffæ*, Seeb.
Canarian Goldcrest.

87. *Regulus ignicapillus* (C. L. Brehm).
Fire-crested Wren.

88. *Regulus maderensis*, Vern. Harc.
Madeiran Goldcrest.

89. *Phylloscopus superciliosus* (Gm.).
Yellow-browed Warbler.

90. *Phylloscopus proregulus* (Pallas).
Pallas's Willow-Warbler.

91. *Phylloscopus tristis*, Blyth.
Siberian Chiffchaff.

92. *Phylloscopus rufus* (Bechst.).
Chiffchaff.

93. *Phylloscopus trochilus* (Linn.).
Willow-Wren.

94. *Phylloscopus sibilatrix* (Bechst.).
Wood-Wren.

95. *Phylloscopus bonellii* (Vieill.).
Bonelli's Warbler.

96. *Phylloscopus viridanus*, Blyth.
Greenish Willow-Warbler.

97. *Phylloscopus nitidus*, Blyth.
Bright Green Willow-Warbler.

98. *Phylloscopus borealis* (Blasius).
Eversmann's Warbler.

99. *Phylloscopus neglectus*, Hume.
Plain Willow-Warbler.

Subfamily ACROCEPHALINÆ.

100. *Hypolais polyglotta* (Vieill.).
Melodious Warbler.

101. *Hypolais icterina* (Vieill.).
Icterine Warbler.

102. *Hypolais olivetorum* (Strickl.).
Olive-tree Warbler.

103. *Hypolais pallida* (Ehr.).
Olivaceous Warbler.

104. *Hypolais opaca* (Licht.).
Western Olivaceous Warbler.

105. *Hypolais languida* (Ehr.).
Upcher's Warbler.

106. *Hypolais caligata* (Licht.).
Booted Warbler.

107. *Hypolais rama* (Sykes).
Sykes's Warbler.

108. *Aëdon galactodes* (Temm.).
Rufous Warbler.

109. *Aëdon familiaris* (Ménétr.).
Grey-backed Warbler.

110. *Acrocephalus agricola* (Jerd.).
Paddy-field Warbler.

111. *Acrocephalus dumetorum*, Blyth.
Blyth's Reed-Warbler.

112. *Acrocephalus streperus* (Vieill.).
Reed-Warbler.

113. *Acrocephalus palustris* (Bechst.).
Marsh-Warbler.

114. *Acrocephalus turdoides* (Meyer).
Great Reed-Warbler.

115. *Acrocephalus stentoreus* (Ehr.).
Clamorous Reed-Warbler.

116. *Acrocephalus aquaticus* (Gmel.).
Aquatic Warbler.

117. *Acrocephalus phragmitis* (Bechst.).
Sedge-Warbler.

118. *Acrocephalus melanopogon* (Temm.).
Moustached Sedge-Warbler.

119. *Locustella nævia* (Bodd.).
Grasshopper-Warbler.

120. *Locustella straminea* (Severtz.).
Eastern Grasshopper-Warbler.

121. *Locustella lanceolata* (Temm.).
Lanceolated Warbler.

122. *Locustella fluviatilis* (Wolf).
River-Warbler.

123. *Locustella luscinioides* (Savi).
Savi's Warbler.

124. *Locustella certhiola* (Pall.).
Pallas's Warbler.

125. *Cettia cettii* (Marm.).'
Cetti's Warbler.

Subfamily DRYMŒCINÆ.

126. *Cisticola cursitans* (Frankl.).
Fantail Warbler.

127. *Drymœca gracilis* (Licht.).
Streaked Wren-Warbler.

128. *Scotocerca inquieta* (Cretz.).
Streaked Scrub-Warbler.

129. *Scotocerca saharæ* (Loche).
Algerian Scrub-Warbler.

Subfamily CRATEROPODINÆ.

130. *Argya fulva* (Desf.).
Algerian Bush-Babbler.

131. *Argya squamiceps* (Cretzschm.).
Palestine Bush-Babbler.

Family ACCENTORIDÆ.

132. *Accentor collaris* (Scop.).
Alpine Accentor.

133. *Accentor montanellus* (Pall.).
Mountain-Accentor.

134. *Accentor fulvescens*, Severtz.
Brown Accentor.

135. *Accentor atrigularis*, Brandt.
Black-throated Accentor.

136. *Accentor modularis* (Linn.).
Hedge-Sparrow.

Family PANURIDÆ.

137. *Panurus biarmicus* (Linn.).
Bearded Titmouse.

Family PARIDÆ.

138. *Acredula rosea* (Blyth).
British Long-tailed Titmouse.

139. *Acredula caudata* (Linn.).
Long-tailed Titmouse.

140. *Acredula irbii*, Sharpe and Dresser.
Irby's Long-tailed Titmouse.

141. *Acredula tephronota* (Günther).
Turkish Long-tailed Titmouse.

142. *Acredula macedonica*, Salvad & Dress.
Macedonian Long-tailed Titmouse.

143. *Acredula caucasica*, Dresser.
Caucasian Long-tailed Titmouse.

144. *Parus major*, Linn.
Great Titmouse.

145. *Parus bokharensis*, Licht.
Bokharan Grey Titmouse.

146. *Parus cinereus*, Vieill.
Indian Grey Titmouse.

147. *Parus ater*, Linn.
European Coal-Titmouse.

148. *Parus britannicus*, Sharpe & Dresser.
British Coal-Titmouse.

149. *Parus ledouci*, Malh.
Algerian Coal-Titmouse.

150. *Parus cypriotes*, Dresser.
Cyprian Coal-Titmouse.

151. *Parus phæonotus*, Blanf.
Persian Coal-Titmouse.

152. *Parus palustris*, Linn.
Marsh-Titmouse.

153. *Parus salicarius*, C. L. Brehm.
Willow Marsh-Titmouse.

154. *Parus borealis*, De Selys.
Northern Marsh-Titmouse.

155. *Parus baicalensis*, Swinhoe.
Siberian Marsh-Titmouse.

156. *Parus lugubris*, Temm.
Sombre Titmouse.

157. *Parus cinctus*, Bodd.
Lapp Titmouse.

158. *Parus cœruleus*, Linn.
Blue Titmouse.

159. *Parus pleskii*, Cab.
Pleske's Blue Titmouse.

160. *Parus ultramarinus*, Bp.
Ultramarine Titmouse.

161. *Parus teneriffæ*, Less.
Canarian Blue Titmouse.

162. *Parus palmensis*, Meade-Waldo.
White-bellied Titmouse.

163. *Parus ombriosus*, Meade-Waldo.
Hierro Titmouse.

164. *Parus cyanus*, Pall.
Azure Titmouse.

165. *Parus cristatus* (Linn.).
Crested Titmouse.

166. *Ægithalus pendulinus* (Linn.).
Penduline Titmouse.

167. *Ægithalus castaneus*, Severtz.
Chestnut-crowned Titmouse.

Family SITTIDÆ.

168. *Sitta europæa*, Linn.
Northern Nuthatch.

169. *Sitta cæsia*, Wolf.
Common Nuthatch.

170. *Sitta neumeyeri*, Michah.
Western Rock-Nuthatch.

171. *Sitta syriaca*, Ehr.
Eastern Rock-Nuthatch.

172. *Sitta krueperi*, von Pelz.
Krüper's Nuthatch.

173. *Sitta whiteheadi*, Sharpe.
Corsican Nuthatch.

Family CARTHIIDÆ.

174. *Certhia familiaris*, Linn.
Tree-Creeper.

175. *Tichodroma muraria* (Linn.).
Wall-Creeper.

Family TROGLODYTIDÆ.

176. *Troglodytes parvulus*, Koch.
 Common Wren.

177. *Troglodytes pallidus*, Hume.
 Pallid Wren.

178. *Troglodytes borealis*, Fisher.
 Northern Wren.

Family MOTACILLIDÆ.

179. *Motacilla alba*, Linn.
 White Wagtail.

180. *Motacilla lugubris*, Temm.
 Pied Wagtail.

181. *Motacilla personata*, Gould.
 Masked Wagtail.

182. *Motacilla citreola*, Pall.
 Yellow-headed Wagtail.

183. *Motacilla melanope*, Pall.
 Grey Wagtail.

184. *Motacilla flava*, Linn.
 Blue-headed Wagtail.

185. *Motacilla viridis*, Gmel.
 Grey-headed Wagtail.

186. *Motacilla melanocephala*, Licht.
 Black-headed Wagtail.

187. *Motacilla xanthophrys*, Sharpe.
 Yellow-browed Wagtail.

188. *Motacilla raii*, Bp.
 Yellow Wagtail.

189. *Anthus pratensis* (Linn.).
Meadow-Pipit.

190. *Anthus bertheloti*, Bolle.
Canarian Pipit.

191. *Anthus gustavi*, Swinh.
Petchora Pipit.

192. *Anthus cervinus* (Pall.).
Red-throated Pipit.

193. *Anthus trivialis* (Linn.).
Tree-pipit.

194. *Anthus campestris* (Linn.).
Tawny Pipit.

195. *Anthus similis* (Jerdon).
Brown Rock-Pipit.

196. *Anthus richardi*, Vieill.
Richard's Pipit.

197. *Anthus ludovicianus* (Gmel.).
Pennsylvanian Pipit.

198. *Anthus spipoletta* (Linn.).
Water-Pipit.

199. *Anthus obscurus* (Lath.).
Rock-Pipit.

Family PYCNONOTIDÆ.

200. *Pycnonotus barbatus* (Desf.).
Dusky Bulbul.

201. *Pycnonotus xanthopygus* (Ehr.).
Palestine Bulbul.

202. *Pycnonotus capensis* (Linn.).
Gold-vented Bulbul.

3

Family ORIOLIDÆ.

203. *Oriolus galbula,* Linn.
 Golden Oriole.

Family LANIIDÆ.

204. *Lanius excubitor,* Linn.
 Great Grey Shrike.

205. *Lanius leucopterus,* Severtz.
 White-winged Shrike.

206. *Lanius funereus,* Menz.
 Eversmann's Shrike.

207. *Lanius elegans,* Swains.
 Pale Shrike.

208. *Lanius grimmi,* Bogd.
 Bogdonoff's Shrike.

209. *Lanius meridionalis,* Temm.
 Southern Grey Shrike.

210. *Lanius algeriensis,* Less.
 Algerian Grey Shrike.

211. *Lanius fallax,* Finsch.
 Finsch's Shrike.

212. *Lanius minor,* Gmel.
 Lesser Grey Shrike.

213. *Lanius raddii,* Dresser.
 Radde's Shrike.

214. *Lanius collurio,* Linn.
 Red-backed Shrike.

215. *Lanius pomeranus,* Sparrm.
 Woodchat Shrike.

216. *Lanius isabellinus*, Ehr.
Isabelline Shrike.

217. *Lanius nubicus*, Licht.
Masked Shrike.

218. *Lanius cucullatus*, Temm.
Hooded Shrike.

Family AMPELIDÆ.

219. *Ampelis garrulus*, Linn.
Waxwing.

Family MUSCICAPIDÆ.

220. *Muscicapa grisola*, Linn.
Spotted Flycatcher.

221. *Muscicapa atricapilla*, Linn.
Pied Flycatcher.

222. *Muscicapa collaris*, Bechst.
White-collared Flycatcher.

223. *Muscicapa semitorquata*, Homeyer.
Caucasian Pied Flycatcher.

224. *Muscicapa parva*, Bechst.
Red-breasted Flycatcher.

Section 2. OSCINES LATIROSTRES.

Family HIRUNDINIDÆ.

225. *Hirundo savignii*, Steph.
Chestnut-bellied Swallow.

226. *Hirundo rustica*, Linn.
Swallow.

227. *Hirundo rufula*, Temm.
   Red-rumped Swallow.

228. *Chelidon urbica* (Linn.).
   Martin.

229. *Cotile riparia* (Linn.).
   Sand-Martin.

230. *Cotile rupestris* (Scop.).
   Crag-Martin.

231. *Cotile obsoleta*, Cab.
   Pale Crag-Martin.

Section 3. **OSCINES CONIROSTRES.**

Family FRINGILLIDÆ.

Subfamily FRINGILLINÆ.

232. *Carduelis elegans*, Steph.
   Goldfinch.

233. *Carduelis caniceps*, Vigors.
   Himalayan Goldfinch.

234. *Carduelis citrinella* (Linn.).
   Citril Finch.

235. *Carduelis spinus* (Linn.).
   Siskin.

236. *Serinus hortulanus*, Koch.
   Serin Finch.

237. *Serinus canonicus*, Dresser.
   Tristram's Serin.

238. *Serinus canarius* (Linn.).
   Canary.

239. *Serinus pusillus* (Pall.).
   Red-fronted Finch.

240. *Ligurinus chloris* (Linn.).
　　Greenfinch.

241. *Coccothraustes vulgaris*, Pall.
　　Hawfinch.

242. *Coccothraustes carneipes*, Hodgson.
　　White-winged Grosbeak.

243. *Passer italiæ* (Vieill.).
　　Italian Sparrow.

244. *Passer domesticus* (Linn.).
　　House-Sparrow.

245. *Passer hispaniolensis*, Temm.
　　Spanish Sparrow.

246. *Passer montanus* (Linn.).
　　Tree-Sparrow.

247. *Passer ammodendri*, Severtz.
　　Saxaul Sparrow.

248. *Passer simplex* (Licht.).
　　Desert-Sparrow.

249. *Petronia stulta* (Gmel.).
　　Rock-Sparrow.

250. *Petronia brachydactyla*, Bp.
　　Desert Rock-Sparrow.

251. *Montifringilla nivalis* (Linn.).
　　Snow-Finch.

252. *Montifringilla alpicola* (Pall.).
　　Eastern Snow-Finch.

253. *Fringilla cœlebs*, Linn.
　　Chaffinch.

254. *Fringilla tintillon*, Webb & Berthel.
　　Azorean Chaffinch.

255. *Fringilla palmæ*, Trist.
　　　Palman Chaffinch.

256. *Fringilla spodiogena*, Bp.
　　　Algerian Chaffinch.

257. *Fringilla montifringilla*, Linn.
　　　Brambling.

258. *Fringilla teydea*, Webb & Berthel.
　　　Teydean Chaffinch.

259. *Linota cannabina* (Linn.).
　　　Linnet.

260. *Linota linaria* (Linn.).
　　　Mealy Redpoll.

261. *Linota rufescens* (Vieill.).
　　　Lesser Redpoll.

262. *Linota exilipes* (Coues).
　　　Coues's Redpoll.

263. *Linota hornemanni*, Holb.
　　　Greenland Redpoll.

264. *Linota flavirostris* (Linn.).
　　　Twite.

265. *Linota brevirostris*, Gould.
　　　Eastern Twite.

　　　　　Subfamily LOXIINÆ.

266. *Carpodacus rubicilla* (Güld.).
　　　Caucasian Rose Finch.

267. *Carpodacus sinaiticus* (Temm.).
　　　Sinaitic Rose Finch.

268. *Carpodacus erythrinus* (Pall.).
　　　Scarlet Grosbeak.

269. *Bucanetes obsoletus* (Licht.).
Desert-Finch.

270. *Bucanetes mongolicus* (Swinhoe).
Mongolian Desert-Finch.

271. *Erythrospiza githaginea* (Licht.).
Trumpeter-Bullfinch.

272. *Erythrospiza sanguinea* (Gould).
Crimson-winged Finch.

273. *Pyrrhula europæa*, Vieill.
Common Bullfinch.

274. *Pyrrhula major*, C. L. Brehm.
Northern Bullfinch.

275. *Pyrrhula murina*, Godm.
Azorean Bullfinch.

276. *Pyrrhula cassini* (Baird).
Cassin's Bullfinch.

277. *Uragus sibiricus* (Pall.).
Siberian Rose-Finch.

278. *Pinicola enucleator* (Linn.).
Pine-Grosbeak.

279. *Loxia pityopsittacus*, Bechst.
Parrot-Crossbill.

280. *Loxia curvirostra*, Linn.
Common Crossbill.

281. *Loxia rubrifasciata* (Brehm).
Red-banded Crossbill.

282. *Loxia leucoptera*, Gmel.
White-winged Crossbill.

283. *Loxia bifasciata* (C. L. Brehm).
Two-barred Crossbill.

Subfamily EMBERIZINÆ.

284. *Emberiza melanocephala*, Scop.
 Black-headed Bunting.

285. *Emberiza luteola*, Sparrm.
 Red-headed Bunting.

286. *Emberiza cinerea*, Strickl.
 Strickland's Bunting.

287. *Emberiza miliaria*, Linn.
 Common Bunting.

288. *Emberiza citrinella*, Linn.
 Yellow Bunting.

289. *Emberiza cirlus*, Linn.
 Cirl Bunting.

290. *Emberiza hortulana*, Linn.
 Ortolan Bunting.

291. *Emberiza huttoni* (Blyth).
 Grey-necked Bunting.

292. *Emberiza chrysophrys*, Pall.
 Yellow-browed Bunting.

293. *Emberiza striolata* (Licht.).
 Striped Bunting.

294. *Emberiza saharæ*, Levaill.
 House-Bunting.

295. *Emberiza cia*, Linn.
 Meadow-Bunting.

296. *Emberiza cioides*, Brandt.
 Siberian Meadow-Bunting.

297. *Emberiza cæsia*, Cretzsch.
 Cretzschmar's Bunting.

298. *Emberiza leucocephala*, Gmel.
Pine-Bunting.

299. *Emberiza aureola*, Pall.
Yellow-breasted Bunting.

300. *Emberiza rustica*, Pall.
Rustic Bunting.

301. *Emberiza pusilla*, Pall.
Little Bunting.

302. *Emberiza schœniclus*, Linn.
Reed-Bunting.

303. *Emberiza pyrrhuloides*, Pall.
Large-billed Reed-Bunting.

304. *Calcarius lapponicus* (Linn.).
Lapland Bunting.

305. *Plectrophenax nivalis* (Linn.).
Snow-Bunting.

Section 4. OSCINES SCUTELLI-
PLANTARES.

Family ALAUDIDÆ.

306. *Certhilauda alaudipes* (Desf.).
Curve-billed Lark.

307. *Certhilauda duponti* (Vieill.).
Dupont's Lark.

308. *Galerita cristata*, Linn.
Crested Lark.

309. *Galerita macrorhyncha* (Trist.).
Tristram's Lark.

310. *Galerita isabellina* (Bp.).
Isabelline Lark.

4

311. *Alauda arvensis*, Linn.
Sky-Lark.

312. *Alauda gulgula*, Franklin.
Indian Sky-Lark.

313. *Alauda arborea*, Linn.
Wood-Lark.

314. *Ammomanes deserti* (Licht.).
Desert-Lark.

315. *Ammomanes cinctura* (Gould).
Gould's Desert-Lark.

316. *Calandrella brachydactyla* (Leisl.).
Short-toed Lark.

317. *Calandrella minor* (Cab.).
Lesser Short-toed Lark.

318. *Calandrella bœtica*, Dresser.
Andalucian Short-toed Lark.

319. *Calandrella pispoletta* (Pall.).
Pallas's Short-toed Lark.

320. *Melanocorypha calandra* (Linn.).
Calandra Lark.

321. *Melanocorypha bimaculata* (Ménétr.).
Eastern Calandra Lark.

322. *Melanocorypha sibirica* (Gmel.).
White-winged Lark.

323. *Melanocorypha yeltoniensis* (Forst.).
Black Lark.

324. *Rhamphocorys clotbey* (Bp.).
Thick-billed Lark.

325. *Otocorys alpestris*, Linn.
Shore-Lark.

326. *Otocorys penicillata* (Gould).
Eastern Shore-Lark.

327. *Otocorys bilopha* (Rüpp.).
Algerian Shore-Lark.

Section 5. **OSCINES CULTRIROSTRES.**

Family STURNIDÆ.

328. *Sturnus vulgaris*, Linn.
Common Starling.

329. *Sturnus unicolor*, De la Marm.
Sardinian Starling.

330. *Sturnus purpurascens*, Gould.
Purple-winged Starling.

331. *Pastor roseus* (Linn.).
Rose-coloured Starling.

Family CORVIDÆ.

332. *Pyrrhocorax graculus* (Linn.).
Red-billed Chough.

333. *Pyrrhocorax alpinus*, Koch.
Alpine Chough.

334. *Podoces panderi*, Fischer.
Pander's Ground-Chough,

335. *Nucifraga caryocatactes* (Linn.).
Nutcracker.

336. *Perisoreus infaustus* (Linn.).
Siberian Jay.

337. *Garrulus glandarius* (Linn.).
Common Jay.

338. *Garrulus hyrcanus* (Blanford).
Persian Jay.

339. *Garrulus minor* (Verreaux).
African Jay.

340. *Garrulus brandti*, Eversm.
Brandt's Jay.

341. *Garrulus atricapillus*, St. Hilaire.
Syrian Jay.

342. *Garrulus krynicki*, Kalenicz.
Black-headed Jay.

343. *Garrulus cervicalis*, Bp.
Algerian Black-headed Jay.

344. *Cyanopica cooki*, Bp.
Azure-winged Magpie.

345. *Pica rustica* (Scop.).
Magpie.

346. *Pica mauritanica*, Malh.
Moorish Magpie.

347. *Corvus monedula*, Linn.
Jackdaw.

348. *Corvus corone*, Linn.
Carrion-Crow.

349. *Corvus cornix*, Linn.
Hooded Crow.

350. *Corvus frugilegus*, Linn.
Rook.

351. *Corvus corax*, Linn.
Raven.

352. *Corvus tingitanus*, Irby.
Tangier Raven.

353. *Corvus affinis*, Rüpp.
Fantailed Raven.

354. *Corvus umbrinus*, Hedenb.
Brown-necked Raven.

Order II. MACROCHIRES.

Family CYPSELIDÆ.

355. *Cypselus apus* (Linn.).
Common Swift.

356. *Cypselus affinis*, J. E. Gray.
White-rumped Swift.

357. *Cypselus pallidus*, Shelley.
Pallid Swift.

358. *Cypselus unicolor*, Jard.
Madeiran Swift.

359. *Cypselus melba* (Linn.).
Alpine Swift.

360. *Acanthyllis caudacuta* (Lath.).
Needle-tailed Swift.

Family CAPRIMULGIDÆ.

361. *Caprimulgus europæus*, Linn.
Common Nightjar.

362. *Caprimulgus ægyptius*, Licht.
Egyptian Nightjar.

363. *Caprimulgus ruficollis*, Temm.
Russet-necked Nightjar.

Order III. PICI.

Family PICIDÆ.

Subfamily PICINÆ.

364. *Picus martius*, Linn.
 Great Black Woodpecker.

365. *Dendrocopus major*, Linn.
 Great Spotted Woodpecker.

366. *Dendrocopus leucopterus*, Salvad.
 White-winged Woodpecker.

367. *Dendrocopus pœlzami*, Bogd.
 Caucasian Spotted Woodpecker.

368. *Dendrocopus numidicus*, Mahl.
 Algerian Pied Woodpecker.

369. *Dendrocopus mauritanus*, Brehm.
 Moorish Pied Woodpecker.

370. *Dendrocopus syriacus*, Ehr.
 Syrian Pied Woodpecker.

371. *Dendrocopus leuconotus*, Bechst.
 White-backed Woodpecker.

372. *Dendrocopus lilfordi*, Sharpe & Dresser.
 Grecian Woodpecker.

373. *Dendrocopus medius*, Linn.
 Middle Spotted Woodpecker.

374. *Dendrocopus sancti-johannis* (Blanf.).
 Persian Woodpecker.

375. *Dendrocopus minor*, Linn.
 Lesser Spotted Woodpecker.

376. *Dendrocopus pipra*, Pall.
 Siberian Lesser Spotted Woodpecker.

377. *Dendrocopus danfordi*, Hargitt.
Turkish Lesser Spotted Woodpecker.

378. *Picoides tridactylus* (Linn.).
Three-toed Woodpecker.

379. *Gecinus viridis* (Linn.).
Green Woodpecker.

380. *Gecinus sharpii*, Saunders.
Spanish Green Woodpecker.

381. *Gecinus vaillanti* (Malh.).
Algerian Green Woodpecker.

382. *Gecinus canus* (Gmel.).
Grey-headed Green Woodpecker.

383. *Gecinus flavirostris*, Zarudny.
Yellow-billed Green Woodpecker.

Subfamily *IYNGINÆ*.

384. *Iynx torquilla*, Linn.
Wryneck.

DESMOGNATHÆ.

Order I. COCCYGES.

Suborder *COCCYGES ANISO-DACTYLI*.

Family ALCEDINIDÆ.

385. *Alcedo ispida*, Linn.
Common Kingfisher.

386. *Ceryle rudis* (Linn.).
Pied Kingfisher.

387. *Halcyon smyrnensis* (Linn.).
Smyrna Kingfisher.

Family CORACIIDÆ.

388. *Coracias garrulus*, Linn.
 Common Roller.

389. *Coracias indicus*, Linn.
 Indian Roller.

Family MEROPIDÆ.

390. *Merops apiaster*, Linn.
 Common Bee-eater.

391. *Merops persicus*, Pall.
 Blue-cheeked Bee-eater.

392. *Merops viridis*, Linn.
 Little Green Bee-eater.

Family UPUPIDÆ.

393. *Upupa epops*, Linn.
 Hoopoe.

Suborder *COCCYGES ZYGODACTYLI*.

Family CUCULIDÆ.

394. *Cuculus canorus*, Linn.
 Cuckoo.

395. *Coccystes glandarius* (Linn.).
 Great Spotted Cuckoo.

396. *Coccyzus americanus* (Linn.).
 Yellow-billed Cuckoo.

397. *Coccyzus erythrophthalmus* (Wils.).
 Black-billed Cuckoo.

Order II. ACCIPITRES.

Suborder *STRIGES*.

Family STRIGIDÆ.

398. *Strix flammea*, Linn.
Barn-Owl.

Family BUBONIDÆ.

399. *Asio otus* (Linn.).
Long-eared Owl.

400. *Asio accipitrinus* (Pall.).
Short-eared Owl.

401. *Asio capensis* (Smith).
Cape Eared Owl.

402. *Syrnium aluco* (Linn.).
Tawny Owl.

403. *Syrnium uralense* (Pall.).
Ural Owl.

404. *Syrnium lapponicum* (Sparrm.).
Lapp Owl.

405. *Nyctea scandiaca* (Linn.).
Snowy Owl.

406. *Surnia ulula* (Linn.).
Hawk-Owl.

407. *Surnia funerea*, Linn.
American Hawk-Owl.

408. *Nyctala tengmalmi* (Gmel.).
Tengmalm's Owl.

409. *Scops giu* (Scop.).
Scops Owl.

410. *Scops brucii* (Hume).
Pallid Scops Owl.

411. *Bubo ignavus*, Forst.
Eagle-Owl.

412. *Bubo ascalaphus*, Savigny.
Egyptian Eagle-Owl.

413. *Glaucidium passerinum* (Linn.).
Pygmy Owl.

414. *Athene noctua* (Retz.).
Little Owl.

415. *Athene glaux* (Savigny).
Southern Little Owl.

416. *Athene bactriana* (Hutton).
Eastern Little Owl.

Suborder ACCIPITRES.

Family VULTURIDÆ.

417. *Gyps fulvus* (Gmel.).
Griffon-Vulture.

418. *Vultur monachus*, Linn.
Black Vulture.

419. *Neophron percnopterus* (Linn.).
Egyptian Vulture.

420. *Gypaëtus barbatus* (Linn.).
Bearded Vulture.

Family FALCONIDÆ.

421. *Circus æruginosus* (Linn.).
 Marsh-Harrier.

422. *Circus cyaneus* (Linn.).
 Hen-Harrier.

423. *Circus cineraceus* (Mont.).
 Montagu's Harrier.

424. *Circus swainsoni*, Smith.
 Pallid Harrier.

425. *Buteo vulgaris*, Leach.
 Common Buzzard.

426. *Buteo desertorum* (Daud.).
 African Buzzard.

427. *Buteo ferox* (Gmel.).
 Long-legged Buzzard.

428. *Archibuteo lagopus* (Gmel.).
 Rough-legged Buzzard.

429. *Aquila pennata* (Gmel.).
 Booted Eagle.

430. *Aquila pomarina*, C. L. Brehm.
 Lesser Spotted Eagle.

431. *Aquila clanga*, Pall.
 Larger Spotted Eagle.

432. *Aquila nipalensis*, Hodgs.
 Steppe-Eagle.

433. *Aquila rapax* (Temm.).
 Tawny Eagle.

434. *Aquila heliaca*, Savigny.
 Imperial Eagle.

435. *Aquila adalberti*, L. Brehm.
Spanish Imperial Eagle.

436. *Aquila chrysaetus* (Linn.).
Golden Eagle.

437. *Haliaëtus albicilla* (Linn.).
White-tailed Eagle.

438. *Haliaëtus leucoryphus*, Pall.
Pallas's White-tailed Eagle.

439. *Circaëtus gallicus* (Gmel.).
Short-toed Eagle.

440. *Nisaëtus fasciatus* (Vieill.).
Bonelli's Eagle.

441. *Astur palumbarius*, Linn.
Goshawk.

442. *Accipiter nisus* (Linn.).
Sparrow-Hawk.

443. *Accipiter badius* (Gmel.).
Shikra.

444. *Accipiter brevipes* (Severtz.).
Levant Sparrow-Hawk.

445. *Milvus ictinus*, Savigny.
Red Kite.

446. *Milvus migrans* (Bodd.).
Black Kite.

447. *Milvus melanotis* (Temm. & Schleg.).
Black-eared Kite.

448. *Milvus ægyptius* (Gmel.).
Arabian Kite.

449. *Elanus cæruleus* (Desf.).
Black-winged Kite.

450. *Pernis apivorus* (Linn.).
Honey-Buzzard.

451. *Falco candicans*, Gmel.
Greenland Falcon.

452. *Falco islandus*, Gmel.
Iceland Falcon.

453. *Falco gyrfalco*, Linn.
Jerfalcon.

454. *Falco peregrinus*, Tunstall.
Peregrine Falcon.

455. *Falco minor*, Bp.
Lesser Peregrine.

456. *Falco barbarus*, Linn.
Barbary Falcon.

457. *Falco feldeggi*, Schlegel.
Lanner.

458. *Falco sacer*, Gmel.
Saker.

459. *Falco milvipes*, Hodgson.
Shanghar Falcon.

460. *Falco subbuteo*, Linn.
Hobby.

461. *Falco eleonoræ*, Gené.
Eleonoran Falcon.

462. *Falco æsalon*, Tunstall.
Merlin.

463. *Falco vespertinus*, Linn.
Red-footed Falcon.

464. *Falco tinnunculus*, Linn.
Kestrel.

465. *Falco cenchris*, Naum.
Lesser Kestrel.

466. *Pandion haliaëtus* (Linn.).
Osprey.

Order III. STEGANOPODES.

Family PELECANIDÆ.

467. *Phalacrocorax carbo* (Linn.).
Cormorant.

468. *Phalacrocorax graculus* (Linn.).
Shag.

469. *Phalacrocorax africanus* (Gmel.).
African Cormorant.

470. *Phalacrocorax pygmæus*, Pall.
Pygmy Cormorant.

471. *Sula bassana* (Linn.).
Gannet.

472. *Pelecanus onocrotalus*, Linn.
Roseate Pelican.

473. *Pelecanus crispus*, Bruch.
Dalmatian Pelican.

Order IV. HERODII.

Family ARDEIDÆ.

474. *Ardea cinerea*, Linn.
Common Heron.

475. *Ardea purpurea*, Linn.
Purple Heron.

476. *Ardea melanocephala*, Childr.
Black-necked Heron.

477. *Ardea alba*, Linn.
Great White Egret.

478. *Ardea garzetta*, Linn.
Lesser Egret.

479. *Ardea bubulcus*, Audouin.
Buff-backed Heron.

480. *Ardea ralloides*, Scop.
Squacco Heron.

481. *Ardetta minuta* (Linn.).
Little Bittern.

482. *Nycticorax griseus* (Linn.).
Night-Heron.

483. *Botaurus stellaris* (Linn.).
Bittern.

484. *Botaurus lentiginosus* (Mont.).
American Bittern.

Family CICONIIDÆ.

485. *Ciconia alba*, Bechst.
White Stork.

486. *Ciconia nigra* (Linn.).
Black Stork.

Family PLATALEIDÆ.

487. *Platalea leucorodia*, Linn.
Spoonbill.

Family IBIDÆ.

488. *Ibis comata* (Rüpp.).
Red-cheeked Ibis.

489. *Ibis æthiopica*, Lath.
Sacred Ibis.

490. *Plegadis falcinellus* (Linn.).
Glossy Ibis.

Family PHŒNICOPTERIDÆ.

491. *Phœnicopterus roseus*, Pall.
Flamingo.

Order V. ANSERES.

Family ANATIDÆ.

492. *Anser cinereus*, Meyer.
Grey Lag-Goose.

493. *Anser segetum* (Gmel.).
Bean-Goose.

494. *Anser brachyrhyncus*, Baill.
Pink-footed Goose.

495. *Anser albifrons* (Scop.).
White-fronted Goose.

496. *Anser erythropus* (Linn.).
Lesser White-fronted Goose.

497. *Bernicla brenta* (Pall.).
Brent Goose.

498. *Bernicla leucopsis* (Bechst.).
Bernacle Goose.

499. *Bernicla ruficollis* (Pall.).
Red-breasted Goose.

500. *Chen albatus* (Cassin).
Cassin's Snow-Goose.

501. *Chen hyberboreus* (Pall.).
Snow-Goose.

502. *Cygnus olor* (Gmel.).
Mute Swan.

503. *Cygnus immutabilis*, Yarr.
Polish Swan.

504. *Cygnus musicus*, Bechst.
Whooper Swan.

505. *Cygnus bewicki*, Yarr.
Bewick's Swan.

506. *Tadorna cornuta* (Gmel.).
Common Sheld-duck.

507. *Casarca rutila* (Pall.).
Ruddy Sheld-duck.

508. *Anas boscas* (Linn.).
Mallard.

509. *Chaulelasmus streperus* (Linn.).
Gadwall.

510. *Spatula clypeata* (Linn.).
Shoveller.

511. *Marmaronetta angustirostris* (Ménétr.).
Marbled Duck.

512. *Nettion crecca* (Linn.).
Common Teal.

513. *Nettion carolinense* (Gmel.).
American Green-winged Teal.

514. *Nettion formosum* (Georgi.).
Baikal Teal.

515. *Querquedula circia* (Linn.).
Garganey Teal.

516. *Querquedula discors* (Linn.).
Blue-winged Teal.

517. *Eunetta falcata* (Georgi.).
Falcated Teal.

518. *Dafila acuta* (Linn.).
Pintail.

519. *Mareca penelope* (Linn.).
Common Wigeon.

520. *Mareca americana* (Gm.).
American Wigeon.

521. *Netta rufina* (Pall.).
Red-crested Pochard.

522. *Fuligula ferina* (Linn.).
Pochard.

523. *Fuligula nyroca. Güld* (Flem.).
White-eyed Duck.

524. *Fuligula marila* (Linn.).
Scaup.

525. *Fuligula cristata* (Leach).
Tufted Duck.

526. *Clangula glaucion* (Linn.).
Goldeneye.

527. *Clangula islandica* (Gmel.).
Barrow's Goldeneye.

528. *Clangula albeola* (Linn.).
Buffel-headed Duck.

529. *Harelda glacialis* (Linn.).
Long-tailed Duck.

530. *Cosmonetta histrionica* (Linn.).
Harlequin Duck.

531. *Somateria mollissima* (Linn.).
Eider Duck.

532. *Somateria spectabilis* (Linn.).
King-Eider.

533. *Heniconetta stelleri* (Pall.).
Steller's Duck.

534. *Œdemia nigra* (Linn.).
Common Scoter.

535. *Œdemia fusca* (Linn.).
Velvet-Scoter.

536. *Œdemia perspicillata* (Linn.).
Surf-Scoter.

537. *Erismatura leucocephala* (Scop.).
White-headed Duck.

538. *Mergus castor*, Linn.
Goosander.

539. *Mergus serrator*, Linn.
Red-breasted Merganser.

540. *Mergus albellus*, Linn.
Smew.

541. *Lophodytes cucullatus* (Linn.).
Hooded Merganser.

SCHIZOGNATHÆ.

Order I. COLUMBÆ.

Family COLUMBIDÆ.

542. *Columba palumbus*, Linn.
Ring-Dove.

543. *Columba casiotis* (Bp.).
Eastern Ring-Dove.

544. *Columba livia*, Bonnat.
Rock-Dove.

545. *Columba œnas*, Linn.
Stock-Dove.

546. *Columba eversmanni*, Bp.
Indian Stock-Dove.

547. *Columba bollii*, Godm.
Bolle's Pigeon.

548. *Columba laurivora*, Webb & Berth.
Canarian Pigeon.

549. *Columba trocaz*, Heinek.
Madeiran Pigeon.

550. *Turtur communis*, Selby.
Turtle-Dove.

551. *Turtur orientalis* (Lath.).
Asiatic Turtle-Dove.

552. *Turtur isabellinus*, Bp.
Isabelline Turtle-Dove.

553. *Turtur risorius* (Linn.).
Collared Turtle-Dove.

554. *Turtur senegalensis* (Linn.).
Egyptian Turtle-Dove.

555. *Turtur cambayensis* (Gm.).
Indian Turtle-Dove.

Family PTEROCLIDÆ.

556. *Pterocles arenarius* (Pall.).
Black-bellied Sand-Grouse.

557. *Pterocles alchata* (Linn.).
Pin-tailed Sand-Grouse.

558. *Pterocles senegallus*, Linn.
Senegal Sand-Grouse.

559. *Pterocles coronatus*, Licht.
Coronetted Sand-Grouse.

560. *Syrrhaptes paradoxus* (Pall.).
Pallas's Sand-Grouse.

Order II. GALLINÆ.

Family PHASIANIDÆ.

561. *Phasianus colchicus*, Linn.
Pheasant.

562. *Phasianus torqatus*, Linn.
Chinese Ring-necked Pheasant.

563. *Phasianus persicus*, Severtz.
Persian Pheasant.

564. *Phasianus principalis*, Sclater.
Murghab Pheasant.

565. *Caccabis saxatilis* (Meyer).
Greek Partridge.

566. *Caccabis chukar* (G. R. Gray).
Chukor Partridge.

567. *Caccabis rufa* (Linn.).
Red-legged Partridge.

568. *Caccabis petrosa* (Gmel.).
Barbary Partridge.

569. *Ammoperdix bonhami* (G. R. Gray).
Seesee Partridge.

570. *Francolinus vulgaris*, Steph.
Francolin.

571. *Francolinus bicalcaratus* (Linn.).
Senegal Francolin.

572. *Perdix cinerea*, Lath.
Partridge.

573. *Coturnix communis*, Bonnat.
Common Quail.

Family TETRAONIDÆ.

574. *Lagopus mutus*, Leach.
Common Ptarmigan.

575. *Lagopus rupestris* (Gmel.).
Rock-Ptarmigan.

576. *Lagopus hemileucurus*, Gould.
Spitsbergen Ptarmigan.

577. *Lagopus albus* (Gmel.).
Willow-Grouse.

578. *Lagopus scoticus* (Lath.).
Red Grouse.

579. *Bonasa betulina* (Scop.).
Hazel-Grouse.

580. *Bonasa griseiventris* (Menzb.).
Menzbier's Hazel-Grouse.

581. *Tetrao tetrix*, Linn.
Black Grouse.

582. *Tetrao mlokosiewiczi*, Taczan.
Georgian Black Grouse.

583. *Tetrao urogallus*, Linn.
Capercaillie.

584. *Tetrao uralensis* (Severtz. & Menzb.).
Ural Capercaillie.

585. *Tetraogallus caucasicus* (Pall.).
Caucasian Snow-Partridge.

586. *Tetraogallus caspius* (Gmel.).
Caspian Snow-Partridge.

Family TURNICIDÆ.

587. *Turnix sylvatica* (Desf.).
Andalucian Hemipode.

Order III. GRALLÆ.

Family RALLIDÆ.

588. *Rallus aquaticus*, Linn.
Water-Rail.

589. *Porzana maruetta* (Leach).
Spotted Crake.

590. *Porzana bailloni* (Vieill.).
Baillon's Crake.

591. *Porzana parva* (Scop.).
Little Crake.

592. *Crex pratensis*, Bechst.
Land-Rail.

593. *Porphyrio cæruleus* (Vandelli).
Purple Gallinule.

594. *Porphyrio poliocephalus* (Lath.).
Indian Gallinule.

595. *Porphyrio smaragdonotus*, Temm.
Green-backed Gallinule.

596. *Porphyrio alleni*, T. R. H. Thomps.
Allen's Gallinule.

597. *Gallinula chloropus* (Linn.).
Moorhen.

598. *Fulica atra*, Linn.
Common Coot.

599. *Fulica cristata*, Gmel.
Crested Coot.

Family GRUIDÆ.

600. *Grus communis*, Bechst.
Common Crane.

601. *Grus virgo* (Linn.).
Demoiselle Crane.

602. *Grus leucogeranus*, Pall.
Siberian Crane.

603. *Grus antigone* (Linn.).
Sarus Crane.

Order IV. LIMICOLÆ.

Family OTIDÆ.

604. *Otis tarda*, Linn.
Great Bustard.

605. *Otis tetrax*, Linn.
Little Bustard.

606. *Otis undulata* (Jacq.).
Houbara Bustard.

607. *Otis macqueeni*, J. E. Gray.
Macqueen's Bustard.

Family ŒDICNEMIDÆ.

608. *Œdicnemus scolopax* (Gmel.).
 Stone-Curlew.

Family GLAREOLIDÆ.

609. *Glareola pratincola*, Linn.
 Common Pratincole.

610. *Glareola melanoptera*, Nordm.
 Nordmann's Pratincole.

Family CHARADRIIDÆ.

611. *Cursorius gallicus* (Gmel.).
 Cream-coloured Courser.

612. *Charadrius pluvialis*, Linn.
 Golden Plover.

613. *Charadrius fulvus*, Gmel.
 Eastern Golden Plover.

614. *Squatarola helvetica* (Linn.).
 Grey Plover.

615. *Ægialitis geoffroyi* (Wagl.).
 Great Sand-Plover.

616. *Ægialitis asiatica* (Pall.).
 Caspian Plover.

617. *Ægialitis cantiana* (Lath.).
 Kentish Plover.

618. *Ægialitis hiaticula* (Linn.).
 Ringed Plover.

619. *Ægialitis curonica* (Gmel.).
 Lesser Ringed Plover.

620. *Ægialitis vocifera* (Linn.).
Killdeer Plover.

621. *Ægialitis pecuaria*, Temm.
Kittlitz's Plover.

622. *Eudromias morinellus* (Linn.).
Dotterel.

623. *Pluvianus ægyptius* (Linn.).
Black-headed Plover.

624. *Chettusia gregaria* (Pall.).
Sociable Plover.

625. *Chettusia leucura* (Licht.).
White-tailed Plover.

626 *Hoplopterus spinosus* (Linn.).
Spur-winged Plover.

627. *Lobivanellus indicus* (Bodd.).
Red-wattled Lapwing.

628. *Vanellus vulgaris*, Bechst.
Lapwing.

629. *Strepsilas interpres* (Linn.).
Turnstone.

630. *Hæmatopus ostralegus*, Linn.
Oystercatcher.

631. *Hæmatopus moquini*, Bp.
African Black Oystercatcher.

Family SCOLOPACIDÆ.

632. *Recurvirostra avocetta*, Linn.
Avocet.

633. *Himantopus candidus*, Bonnat.
Stilt.

634. *Phalaropus hyperboreus* (Linn.).
Red-necked Phalarope.

635. *Phalaropus fulicarius*, Linn.
Grey Phalarope.

636. *Scolopax rusticula*, Linn.
Woodcock.

637. *Gallinago major* (Gmel.).
Great Snipe.

638. *Gallinago cœlestis* (Frenzel).
Common Snipe.

639. *Gallinago gallinula* (Linn.).
Jack Snipe.

640. *Limicola platyrhyncha* (Temm.).
Broad-billed Sandpiper.

641. *Tringa maculata*, Vieill.
Pectoral Sandpiper.

642. *Tringa acuminata* (Horsf.).
Sharp-tailed Sandpiper.

643. *Tringa fuscicollis*, Vieill.
Bonaparte's Sandpiper.

644. *Tringa alpina*, Linn.
Dunlin.

645. *Tringa minuta*, Leisl.
Little Stint.

646. *Tringa temmincki*, Leisl.
Temminck's Stint.

647. *Tringa minutilla*, Vieill.
American Stint.

648. *Tringa subarquata* (Güld.).
Curlew-Sandpiper.

649. *Tringa striata*, Linn.
Purple Sandpiper.

650. *Tringa canutus*, Linn.
Knot.

651. *Machetes pugnax* (Linn.).
Ruff.

652. *Calidris arenaria* (Linn.).
Sanderling.

653. *Tryngites rufescens* (Vieill.).
Buff-breasted Sandpiper.

654. *Bartramia longicauda* (Bechst.).
Bartram's Sandpiper.

655. *Totanus hypoleucus* (Linn.).
Common Sandpiper.

656. *Totanus macularius* (Linn.).
Spotted Sandpiper.

657. *Totanus ochropus* (Linn.).
Green Sandpiper.

658. *Totanus glareola* (Linn.).
Wood-Sandpiper.

659. *Totanus solitarius* (Wils.).
Solitary Sandpiper.

660. *Totanus stagnatilis* (Bechst.).
Marsh-Sandpiper.

661. *Totanus calidris* (Linn.).
Common Redshank.

662. *Totanus fuscus* (Linn.).
Spotted Redshank.

663. *Totanus canescens* (Gmel.).
Greenshank.

664. *Totanus flavipes* (Gmel.).
Yellowshank.

665. *Macroramphus griseus* (Gmel.).
Red-breasted Snipe.

666. *Terekia cinerea* (Guild).
Terek Sandpiper.

667. *Limosa lapponica* (Linn.).
Bar-tailed Godwit.

668. *Limosa ægocephala* (Linn.).
Black-tailed Godwit.

669. *Numenius borealis* (Forst.).
Eskimo Curlew.

670. *Numenius phæopus* (Linn.).
Whimbrel.

670a. *Numenius hudsonicus*, Lath.
Hudsonian Curlew.

671. *Numenius tenuirostris*, Vieill.
Slender-billed Curlew.

672. *Numenius arquata* (Linn.).
Common Curlew.

Order V. GAVIÆ.

Family LARIDÆ.

Subfamily STERNINÆ.

673. *Sterna macrura*, Naum.
Arctic Tern.

674. *Sterna fluviatilis*, Naum.
Common Tern.

675. *Sterna dougalli*, Mont.
Roseate Tern.

676. *Sterna minuta,* Linn.
Little Tern.

677. *Sterna media,* Cretzchm.
Allied Tern.

678. *Sterna caspia,* Pall.
Caspian Tern.

679. *Sterna maxima,* Bodd.
Royal Tern.

680. *Sterna anglica,* Mont.
Gull-billed Tern.

681. *Sterna cantiaca,* Gmel.
Sandwich Tern.

682. *Sterna fuliginosa,* Gmel.
Sooty Tern.

683. *Anous stolidus,* Linn.
Noddy.

684. *Hydrochelidon hybrida* (Pall.).
Whiskered Tern.

685. *Hydrochelidon leucoptera* (Schinz).
White-winged Black Tern.

686. *Hydrochelidon nigra* (Linn.).
Black Tern.

Subfamily LARINÆ.

687. *Xema sabinii* (Sabine).
Sabine's Gull.

688. *Rhodostethia rosea* (Macgill.).
Cuneate-tailed Gull.

689. *Pagophila eburnea* (Phipps).
Ivory Gull.

690. *Larus ridibundus*, Linn.
Black-headed Gull.

691. *Larus melanocephalus*, Natt.
Adriatic Gull.

692. *Larus philadelphia* (Ord.).
Bonaparte's Gull.

693. *Larus ichthyaëtus*, Pall.
Great Black-headed Gull.

694. *Larus minutus*, Pall.
Little Gull.

695. *Larus canus*, Linn.
Common Gull.

696. *Larus gelastes*, Licht.
Slender-billed Gull.

697. *Larus audouini*, Payr.
Audouin's Gull.

698. *Larus argentatus*, Gmel.
Herring-Gull.

699. *Larus cachinnans*, Pall.
Yellow-legged Herring-Gull.

700. *Larus affinis*, Reinh.
Siberian River-Gull.

701. *Larus fuscus*, Linn.
Lesser Black-backed Gull.

702. *Larus marinus*, Linn.
Great Black-backed Gull.

703. *Larus glaucus*, Fabr.
Glaucous Gull.

704. *Larus leucopterus*, Faber.
Iceland Gull.

705. *Rissa tridactyla* (Linn.).
Kittiwake.

Subfamily STERCORARIINÆ.

706. *Stercorarius catarrhactes*, Linn.
Great Skua.

707. *Stercorarius pomatorhinus* (Temm.).
Pomatorhine Skua.

708. *Stercorarius crepidatus* (Banks).
Richardson's Skua.

709. *Stercorarius parasitius* (Linn.).
Buffon's Skua.

Order VI. TURBINARES.

Family PROCELLARIIDÆ.

710. *Procellaria pelagica*, Linn.
Storm-Petrel.

711. *Oceanodroma leucorrhoa* (Vieill.).
Leach's Petrel.

712. *Oceanodroma castro*, Harcourt.
Harcourt's Petrel.

713. *Oceanites oceanicus* (Kuhl.).
Wilson's Petrel.

714. *Pelagodroma marina* (Lath.).
Frigate Petrel.

715. *Puffinus kuhli* (Boie).
Mediterranean Great Shearwater.

716. *Puffinus anglorum* (Temm.).
Manx Shearwater.

717. *Puffinus griseus* (Gmel.).
Sooty Shearwater.

718. *Puffinus obscurus* (Gmel.).
Dusky Shearwater.

719. *Puffinus assimilis*, Gould.
Gould's Shearwater.

720. *Puffinus gravis* (O'Reilly).
Great Shearwater.

721. *Fulmarus glacialis*, Linn.
Fulmar.

722. *Œstrelata hæsitata* (Kuhl).
Capped Petrel.

723. *Œstrelata brevipes* (Peale).
Short-toed Petrel.

724. *Œstrelata mollis*, Gould.
Soft-plumaged Petrel.

725. *Bulweria bulweri* (Jard & Selby).
Bulwer's Petrel.

Family DIOMEDEA.

726. *Diomedia melanophrys*.
Black-browed Albatross.

Order VII. ALCÆ.

Family ALCIDÆ.

727. *Alca torda*, Linn.
Razorbill.

728. *Alca impennis*, Linn.
Garefowl.

729. *Lomvia troile* (Linn.).
Common Guillemot.

730. *Lomvia bruennichi* (Sabine).
Brünnich's Guillemot.

731. *Uria grylle* (Linn.).
Black Guillemot.

732. *Uria mandti*, Licht.
Spitsbergen Guillemot.

733. *Mergulus alle* (Linn.).
Little Auk.

734. *Fratercula arctica* (Linn.).
Puffin.

Order VIII. PYGOPODES.

Family COLYMBIDÆ.

735. *Colymbus glacialis*, Linn.
Great Northern Diver.

736. *Colymbus adamsi*, G. R. Gray.
Yellow-billed Diver.

737. *Colymbus arcticus*, Linn.
Black-throated diver.

738. *Colymbus septentrionalis*, Linn.
Red-throated Diver.

Family PODICIPEDIDÆ.

739. *Podicipes cristatus* (Linn.).
Great Crested Grebe.

740. *Podicipes griseigena* (Bodd.)
Red-necked Grebe.

741. *Podicipes auritus* (Linn.).
Sclavonian Grebe.

742. *Podicipes nigricollis* (C. L. Brehm).
    Black-necked Grebe.

743. *Podicipes fluviatilis* (Tunstall).
    Little Grebe.

LIST of Birds recorded as having occurred in the Western Palæarctic Area, but which are not considered entitled to admission.

1. *Turdus pallidus*, Gm.
    Pallid Thrush.

2. *Turdus olivaceus*, Linn.
    Olive Thrush.

3. *Turdus migratorius*, Linn.
    American Robin.

4. *Oreocincla dauma* (Lath.).
    Himalayan Ground-Thrush.

5. *Harporhynchus rufus* (Linn.).
    Brown Thrasher.

6. *Galeoscoptes carolinensis* (Linn.).
    Cat-Bird.

7. *Cinclus pallasi*, Temm.
    Pallas's Dipper.

8. *Regulus calendula* (Linn.).
    Ruby-crowned Kinglet.

9. *Phylloscopus coronatus* (Temm.).
    Grey-legged Willow-Wren.

10. *Dendrœca virens* (Gm.).
    Black-throated Green Warbler.

11. *Parus minor* (Schlegel).
    Japanese Grey Titmouse.

12. *Lophophanes bicolor* (Linn.).
    Tufted Titmouse.

13. *Lanius ludovicianus*, Linn.
    American Grey Shrike.

14. *Vireo olivaceus* (Linn.).
    Red-eyed Flycatcher.

15. *Ampelis cedrorum* (Vieill.).
    Cedar-Bird.

16. *Tachycineta bicolor* (Vieill.).
    White-bellied Swallow.

17. *Progne subis* (Linn.).
    Purple Martin.

18. *Chelidon cashmiriensis*, Gould.
    Kashmir Martin.

19. *Cyanospiza ciris* (Linn.).
    Painted Bunting.

20. *Serinus icterus* (Bonn. & Vieill.).
    Golden-rumped Seedeater.

21. *Carpodacus roseus* (Pall.).
    Rose Finch.

22. *Carpodacus rhodochlamys* (Brandt.).
    Red-mantled Rose Finch.

23. *Zonotrichia albicollis* (Gm.).
    White-throated Sparrow.

24. *Dolichonyx oryzivorus* (Linn.).
    Bobolink.

25. *Junco hiemalis* (Linn.).
    Black Snow-Bird.

26. *Emberiza fucata*, Pall.
    Grey-headed Bunting.

27. *Agelæus phœniceus* (Linn.).
    Red-winged Starling.

28. *Scolecophagus ferrugineus* (Gmel.).
    Rusty Grackle.

29. *Sturnella magna* (Linn.).
    Meadow-Starling.

30. *Caprimulgus nubicus*, Licht.
    Nubian Goatsucker.

31. *Dendrocopus villosus*, Forst.
    Hairy Woodpecker.

32. *Dendrocopus pubescens*, Linn.
    Downy Woodpecker.

33. *Colaptes auratus* (Linn.).
    Golden-winged Woodpecker.

34. *Ceryle alcyon* (Linn.).
    Belted Kingfisher.

35. *Coracias abyssinicus*, Gm.
    Abyssinian Roller.

36. *Merops philippinus*, Linn.
    Blue-tailed Bee-eater.

37. *Cuculus himalayanus*, Blyth.
    Himalayan Cuckoo.

38. *Syrnium nebulosum* (Forst.).
    Barred Owl.

39. *Nyctala acadica* (Gmel.).
    Saw-whet Owl.

40. *Scops asio* (Linn.).
    American Screech-Owl.

41. *Bubo sibiricus* (Schl. & Susem).
    Siberian Eagle-Owl.

42. *Otogyps auricularis* (Daud.).
    Northern Sociable Vulture.

43. *Buteo borealis* (Gmel.).
    Red-tailed Buzzard.

44. *Buteo lineatus* (Gmel.).
   Red-shouldered Buzzard.

45. *Archibuteo sancti-johannis* (Gmel.).
   American Rough-legged Buzzard.

46. *Astur atricapillus* (Wils.).
   American Goshawk.

47. *Accipiter granti*, Sharpe.
   Madeiran Sparrow-Hawk.

48. *Melierax gabar* (Daud.).
   Red-billed Hawk.

49. *Elanoides furcatus* (Linn.).
   Swallow-tailed Kite.

50. *Falco concolor*, Temm.
   Sooty Falcon.

51. *Bubulcus coromandus* (Bodd.).
   Cattle-Egret.

52. *Ardeola sturmi* (Wagl.).
   African Dwarf-Bittern.

53. *Butorides virescens* (Linn.).
   Green Heron.

54. *Plectropterus gambensis* (Linn.).
   Spur-winged Goose.

55. *Chenalopex ægyptiacus* (Linn.),
   Egyptian Goose.

56. *Anser cygnoides* (Linn.).
   Chinese Goose.

57. *Anser indicus* (Lath.).
   Bar-headed Goose.

58. *Bernicla canadensis* (Linn.).
   Canada Goose.

59. *Bernicla canagica* (Sevast.).
   Emperor Goose.

60. *Cygnus americanus*, Sharpless.
    American Swan.

61. *Cygnus buccinator*, Richardson.
    Trumpeter-Swan.

62. *Cairina moschata* (Linn.).
    Muscovy Duck.

63. *Dendrocygna javanica* (Horsf.).
    Tree-Duck.

64. *Æx sponsa* (Linn.).
    Summer-Duck.

65. *Fuligula collaris* (Donov.).
    Ring-necked Duck.

66. *Fuligula affinis*, Eyton.
    Lesser Scaup.

67. *Ectopistes migratorius* (Linn.).
    Passenger-Pigeon.

68. *Ortyx virginianus* (Linn.).
    Virginian Quail.

69. *Porzana carolina* (Linn.).
    Carolina Rail.

70. *Porphyrio martinicus* (Linn.).
    Martinique Gallinule.

71. *Balearica pavonina* (Linn.).
    Balearic Crane.

72. *Œdicnemus senegalensis*, Sw.
    Senegal Thick-knee.

73. *Charadrius dominicus*, Mull.
    American Golden Plover.

74. *Ægialitis mongolica* (Pall.).
    Mongolian Plover.

75. *Gallinago wilsoni* (Temm.).
    Wilson's Snipe.

76. *Symphemia semipalmata* (Gm.).
    Willet.

77. *Sterna bergi*, Licht.
    Swift Tern.

78. *Sterna anæstheta*, Scop.
    Lesser Sooty Tern.

79. *Larus atricilla*, Linn.
    Laughing Gull.

80. *Larus leucopthalmus*, Licht.
    White-eyed Gull.

81. *Daption capense* (Linn.).
    Cape Petrel.

82. *Prion ariel*, Gould.
    Fairy Prion.

83. *Phaëton æthereus*, Linn.
    Tropic-Bird.

84. *Diomedia exulans*, Linn.
    Wandering Albatross.

85. *Thalassogeron culminatus* (Gould).
    Yellow-nosed Albatross.

86. *Tachypetes aquilus*, Linn.
    Frigate-Bird.

87. *Ossifraga gigantea* (Gm.).
    Giant Petrel.

88. *Mormon corniculata*, Naum.
    Horned Puffin.

89. *Podilymbus podicipes* (Linn.).
    Pied-billed Grebe.

www.ingramcontent.com/pod-product-compliance
Lightning Source LLC
Chambersburg PA
CBHW031345160426
43196CB00007B/735